LEARN AND GROW PRESENTS

MY PIRATE FRIENDS

Addition 1-10 Practice Workbook

TKTCOLLECTION PUBLISHING

FIRST EDITION SEPTEMBER 2022

ISBN: 978-1-959247-04-3

Look Up Charts

ADDITION TABLE

1 + 0 = 1	2 + 0 = 2	3 + 0 = 3	4 + 0 = 4	5 + 0 = 5
1 + 1 = 2	2 + 1 = 3	3 + 1 = 4	4 + 1 = 5	5 + 1 = 6
1 + 2 = 3	2 + 2 = 4	3 + 2 = 5	4 + 2 = 6	5 + 2 = 7
1 + 3 = 4	2 + 3 = 5	3 + 3 = 6	4 + 3 = 7	5 + 3 = 8
1 + 4 = 5	2 + 4 = 6	3 + 4 = 7	4 + 4 = 8	5 + 4 = 9
1 + 5 = 6	2 + 5 = 7	3 + 5 = 8	4 + 5 = 9	5 + 5 = 10
1 + 6 = 7	2 + 6 = 8	3 + 6 = 9	4 + 6 = 10	5 + 6 = 11
1 + 7 = 8	2 + 7 = 9	3 + 7 = 10	4 + 7 = 11	5 + 7 = 12
1 + 8 = 9	2 + 8 = 10	3 + 8 = 11	4 + 8 = 12	5 + 8 = 13
1 + 9 = 10	2 + 9 = 11	3 + 9 = 12	4 + 9 = 13	5 + 9 = 14
1 + 10 = 11	2 + 10 = 12	3 + 10 = 13	4 + 10 = 14	5 + 10 = 15

6 + 0 = 6	7 + 0 = 7	8 + 0 = 8	9 + 0 = 9	10 + 0 = 10
6 + 1 = 7	7 + 1 = 8	8 + 1 = 9	9 + 1 = 10	10 + 1 = 11
6 + 2 = 8	7 + 2 = 9	8 + 2 = 10	9 + 2 = 11	10 + 2 = 12
6 + 3 = 9	7 + 3 = 10	8 + 3 = 11	9 + 3 = 12	10 + 3 = 13
6 + 4 = 10	7 + 4 = 11	8 + 4 = 12	9 + 4 = 13	10 + 4 = 14
6 + 5 = 11	7 + 5 = 12	8 + 5 = 13	9 + 5 = 14	10 + 5 = 15
6 + 6 = 12	7 + 6 = 13	8 + 6 = 14	9 + 6 = 15	10 + 6 = 16
6 + 7 = 13	7 + 7 = 14	8 + 7 = 15	9 + 7 = 16	10 + 7 = 17
6 + 8 = 14	7 + 8 = 15	8 + 8 = 16	9 + 8 = 17	10 + 8 = 18
6 + 9 = 15	7 + 9 = 16	8 + 9 = 17	9 + 9 = 18	10 + 9 = 19
6 + 10 = 16	7 + 10 = 17	8 + 10 = 18	9 + 10 = 19	10 + 10 = 20

The highlighted numbers will not be in this book, but are handy to have!

This page is called a "Reference Page."

If you get stuck at any time, You can return to this page to look for what a word means or a solution.

Addition 1-10

Section 1

Fill in the Missing Number

Fill in the Missing Number to complete the Number Sentence.

Example #1

$2 + \square = 5$

Answer

$2 + \boxed{3} = 5$

The Missing Number can be in different parts of the Number Sentence.

Example #2

$\square + 3 = 5$

Answer

$\boxed{2} + 3 = 5$

Find the Sum

The answer to an Addition Problem is the Sum

Example

$2 + 3 = \square$

Answer

$2 + 3 = \boxed{5}$

Fill in the Missing Numbers to Complete the Number Sentance.

1) $1 + \square = 3$

2) $4 + \square = 9$

3) $1 + \square = 4$

4) $3 + \square = 8$

5) $3 + \square = 4$

6) $2 + \square = 5$

7) $3 + \square = 4$

8) $3 + \square = 7$

9) $4 + \square = 6$

10) $2 + \square = 5$

11) $3 + \square = 6$

12) $2 + \square = 6$

Fill in the Missing Numbers to Complete the Number Sentance.

1) 5 + ☐ = 9

2) 3 + ☐ = 8

3) 5 + ☐ = 10

4) 1 + ☐ = 6

5) 3 + ☐ = 4

6) 4 + ☐ = 8

7) 3 + ☐ = 7

8) 1 + ☐ = 6

9) 4 + ☐ = 5

10) 4 + ☐ = 6

11) 5 + ☐ = 8

12) 5 + ☐ = 9

Fill in the Missing Numbers to Complete the Number Sentance.

1) 5 + ☐ = 8

2) 2 + ☐ = 7

3) 5 + ☐ = 7

4) 1 + ☐ = 3

5) 4 + ☐ = 9

6) 5 + ☐ = 9

7) 2 + ☐ = 5

8) 5 + ☐ = 9

9) 4 + ☐ = 7

10) 3 + ☐ = 6

11) 5 + ☐ = 8

12) 2 + ☐ = 7

Fill in the Missing Numbers to Complete the Number Sentance.

1) 1 + ☐ = 5

2) 5 + ☐ = 8

3) 4 + ☐ = 7

4) 1 + ☐ = 2

5) 2 + ☐ = 6

6) 1 + ☐ = 3

7) 1 + ☐ = 2

8) 2 + ☐ = 3

9) 2 + ☐ = 5

10) 5 + ☐ = 8

11) 1 + ☐ = 3

12) 5 + ☐ = 9

Fill in the Missing Numbers to Complete the Number Sentance.

1) $1 + \square = 5$

2) $5 + \square = 10$

3) $3 + \square = 4$

4) $2 + \square = 3$

5) $4 + \square = 7$

6) $5 + \square = 6$

7) $5 + \square = 10$

8) $4 + \square = 6$

9) $1 + \square = 4$

10) $1 + \square = 4$

11) $4 + \square = 6$

12) $5 + \square = 7$

Fill in the Missing Numbers to Complete the Number Sentance.

1) 5 + ☐ = 10

2) 2 + ☐ = 7

3) 3 + ☐ = 8

4) 2 + ☐ = 7

5) 4 + ☐ = 7

6) 2 + ☐ = 4

7) 5 + ☐ = 6

8) 3 + ☐ = 4

9) 3 + ☐ = 4

10) 4 + ☐ = 7

11) 3 + ☐ = 8

12) 3 + ☐ = 5

Fill in the Missing Numbers to Complete the Number Sentance.

1) $3 + \square = 7$

2) $3 + \square = 4$

3) $1 + \square = 2$

4) $2 + \square = 5$

5) $4 + \square = 8$

6) $3 + \square = 8$

7) $4 + \square = 8$

8) $3 + \square = 8$

9) $4 + \square = 5$

10) $3 + \square = 4$

11) $5 + \square = 6$

12) $1 + \square = 2$

7

Fill in the Missing Numbers to Complete the Number Sentance.

1) 5 + ___ = 10

2) 5 + ___ = 8

3) 1 + ___ = 4

4) 5 + ___ = 7

5) 2 + ___ = 6

6) 1 + ___ = 5

7) 2 + ___ = 7

8) 5 + ___ = 7

9) 2 + ___ = 6

10) 4 + ___ = 7

11) 1 + ___ = 2

12) 4 + ___ = 8

Fill in the Missing Numbers to Complete the Number Sentance.

1) $\square + 5 = 7$ 2) $\square + 4 = 9$

3) $\square + 5 = 7$ 4) $\square + 2 = 3$

5) $\square + 5 = 7$ 6) $\square + 1 = 5$

7) $\square + 5 = 9$ 8) $\square + 3 = 7$

9) $\square + 3 = 6$ 10) $\square + 2 = 5$

11) $\square + 3 = 8$ 12) $\square + 4 = 6$

Fill in the Missing Numbers to Complete the Number Sentance.

1) $\square + 4 = 9$

2) $\square + 1 = 4$

3) $\square + 3 = 5$

4) $\square + 1 = 5$

5) $\square + 1 = 5$

6) $\square + 1 = 3$

7) $\square + 1 = 4$

8) $\square + 2 = 7$

9) $\square + 5 = 8$

10) $\square + 5 = 9$

11) $\square + 5 = 8$

12) $\square + 3 = 7$

Fill in the Missing Numbers to Complete the Number Sentance.

1) ☐ + 3 = 4

2) ☐ + 4 = 8

3) ☐ + 1 = 4

4) ☐ + 5 = 8

5) ☐ + 5 = 6

6) ☐ + 5 = 6

7) ☐ + 1 = 2

8) ☐ + 3 = 6

9) ☐ + 2 = 4

10) ☐ + 1 = 3

11) ☐ + 4 = 9

12) ☐ + 1 = 6

Fill in the Missing Numbers to Complete the Number Sentance.

1) ____ + 3 = 5

2) ____ + 4 = 8

3) ____ + 2 = 6

4) ____ + 5 = 6

5) ____ + 2 = 6

6) ____ + 1 = 6

7) ____ + 5 = 6

8) ____ + 4 = 8

9) ____ + 2 = 7

10) ____ + 4 = 7

11) ____ + 1 = 6

12) ____ + 5 = 9

Fill in the Missing Numbers to Complete the Number Sentance.

1) $\square + 2 = 6$

2) $\square + 2 = 5$

3) $\square + 5 = 7$

4) $\square + 3 = 8$

5) $\square + 1 = 5$

6) $\square + 3 = 6$

7) $\square + 1 = 4$

8) $\square + 3 = 6$

9) $\square + 5 = 10$

10) $\square + 2 = 6$

11) $\square + 3 = 8$

12) $\square + 3 = 6$

Fill in the Missing Numbers to Complete the Number Sentance.

1) ☐ + 5 = 6

2) ☐ + 1 = 6

3) ☐ + 4 = 7

4) ☐ + 1 = 5

5) ☐ + 1 = 6

6) ☐ + 5 = 7

7) ☐ + 1 = 3

8) ☐ + 1 = 5

9) ☐ + 5 = 9

10) ☐ + 5 = 10

11) ☐ + 2 = 6

12) ☐ + 2 = 5

Fill in the Missing Numbers to Complete the Number Sentance.

1) $\square + 2 = 4$

2) $\square + 3 = 6$

3) $\square + 1 = 2$

4) $\square + 2 = 3$

5) $\square + 3 = 7$

6) $\square + 1 = 2$

7) $\square + 2 = 7$

8) $\square + 4 = 9$

9) $\square + 3 = 8$

10) $\square + 5 = 9$

11) $\square + 2 = 5$

12) $\square + 2 = 6$

Fill in the Missing Numbers to Complete the Number Sentance.

1) $\square + 2 = 5$

2) $\square + 5 = 6$

3) $\square + 1 = 6$

4) $\square + 4 = 6$

5) $\square + 4 = 6$

6) $\square + 4 = 9$

7) $\square + 1 = 4$

8) $\square + 3 = 4$

9) $\square + 2 = 6$

10) $\square + 2 = 4$

11) $\square + 2 = 7$

12) $\square + 1 = 4$

Complete the Number Sentences to Find the Sum.

1) 5 + 1 = ☐

2) 5 + 4 = ☐

3) 5 + 4 = ☐

4) 3 + 2 = ☐

5) 2 + 5 = ☐

6) 1 + 4 = ☐

7) 2 + 4 = ☐

8) 4 + 2 = ☐

9) 1 + 4 = ☐

10) 5 + 5 = ☐

11) 3 + 1 = ☐

12) 3 + 5 = ☐

Complete the Number Sentences to Find the Sum.

1) 4 + 4 = ☐ 2) 4 + 3 = ☐

3) 4 + 4 = ☐ 4) 2 + 3 = ☐

5) 2 + 2 = ☐ 6) 4 + 4 = ☐

7) 3 + 5 = ☐ 8) 3 + 3 = ☐

9) 1 + 5 = ☐ 10) 2 + 1 = ☐

11) 4 + 5 = ☐ 12) 3 + 2 = ☐

Complete the Number Sentences to Find the Sum.

1) 2 + 4 = ☐ 2) 5 + 1 = ☐

3) 5 + 5 = ☐ 4) 2 + 2 = ☐

5) 4 + 4 = ☐ 6) 2 + 1 = ☐

7) 3 + 5 = ☐ 8) 2 + 4 = ☐

9) 4 + 2 = ☐ 10) 5 + 4 = ☐

11) 4 + 3 = ☐ 12) 4 + 4 = ☐

Complete the Number Sentences to Find the Sum.

1) 5 + 4 = ☐

2) 1 + 4 = ☐

3) 5 + 4 = ☐

4) 5 + 4 = ☐

5) 3 + 2 = ☐

6) 4 + 2 = ☐

7) 2 + 5 = ☐

8) 4 + 5 = ☐

9) 5 + 3 = ☐

10) 1 + 4 = ☐

11) 1 + 4 = ☐

12) 5 + 5 = ☐

Complete the Number Sentences to Find the Sum.

1) 4 + 1 = ☐

2) 1 + 4 = ☐

3) 4 + 4 = ☐

4) 4 + 3 = ☐

5) 1 + 1 = ☐

6) 5 + 3 = ☐

7) 2 + 4 = ☐

8) 3 + 3 = ☐

9) 2 + 2 = ☐

10) 2 + 3 = ☐

11) 1 + 4 = ☐

12) 5 + 4 = ☐

Complete the Number Sentences to Find the Sum.

1) 1 + 2 = ☐	2) 3 + 5 = ☐
3) 5 + 2 = ☐	4) 1 + 2 = ☐
5) 5 + 1 = ☐	6) 5 + 2 = ☐
7) 2 + 4 = ☐	8) 4 + 4 = ☐
9) 4 + 4 = ☐	10) 3 + 3 = ☐
11) 3 + 3 = ☐	12) 3 + 3 = ☐

Complete the Number Sentences to Find the Sum.

1) 3 + 5 = ☐

2) 3 + 3 = ☐

3) 4 + 1 = ☐

4) 4 + 2 = ☐

5) 4 + 2 = ☐

6) 5 + 1 = ☐

7) 1 + 1 = ☐

8) 1 + 4 = ☐

9) 4 + 2 = ☐

10) 4 + 5 = ☐

11) 5 + 5 = ☐

12) 4 + 5 = ☐

Complete the Number Sentences to Find the Sum.

1) 1 + 1 = ☐

2) 5 + 3 = ☐

3) 5 + 3 = ☐

4) 1 + 2 = ☐

5) 5 + 1 = ☐

6) 1 + 5 = ☐

7) 3 + 3 = ☐

8) 1 + 4 = ☐

9) 5 + 2 = ☐

10) 2 + 2 = ☐

11) 2 + 3 = ☐

12) 3 + 5 = ☐

Section 2

These Number Sentences look different, but they are just written differently.

Example #1	Example #2
2 + 3 = 5	$\begin{array}{r} 2 \\ + \, 3 \\ \hline 5 \\ \hline \end{array}$

Here are the Numbers

2 + 3 = 5 $\begin{array}{r} 2 \\ 3 \\ \hline 5 \\ \hline \end{array}$

Here are the Symbols

2 + 3 = 5 $\begin{array}{r} 2 \\ + \, 3 \\ \hline \\ \hline \end{array}$

Section 2

Fill in the Missing Number

Fill in the Missing Number to complete the Number Sentence.

Example #1

$$\begin{array}{r} \square \\ +\ 3 \\ \hline 5 \end{array}$$

Answer

$$\begin{array}{r} 2 \\ +\ 3 \\ \hline 5 \end{array}$$

The Missing Number can be in different parts of the Number Sentence.

Example #2

$$\begin{array}{r} 2 \\ +\ \square \\ \hline 5 \end{array}$$

Answer

$$\begin{array}{r} 2 \\ +\ 3 \\ \hline 5 \end{array}$$

Find the Sum

The answer to an Addition Problem is the Sum

Example

$$\begin{array}{r} 2 \\ +\ 3 \\ \hline \square \end{array}$$

Answer

$$\begin{array}{r} 2 \\ +\ 3 \\ \hline 5 \end{array}$$

Fill in the Missing Number.

1) $\square + 4 = 8$	2) $\square + 2 = 5$	3) $\square + 1 = 6$
4) $\square + 2 = 7$	5) $\square + 1 = 5$	6) $\square + 2 = 6$
7) $\square + 1 = 5$	8) $\square + 1 = 2$	9) $\square + 2 = 3$
10) $\square + 5 = 6$	11) $\square + 1 = 6$	12) $\square + 1 = 2$
13) $\square + 5 = 9$	14) $\square + 1 = 5$	15) $\square + 2 = 7$

Fill in the Missing Number.

1) ☐ + 4 6	2) ☐ + 2 7	3) ☐ + 4 7
4) ☐ + 2 5	5) ☐ + 5 7	6) ☐ + 1 5
7) ☐ + 3 4	8) ☐ + 5 8	9) ☐ + 1 3
10) ☐ + 4 8	11) ☐ + 2 3	12) ☐ + 4 9
13) ☐ + 3 4	14) ☐ + 3 5	15) ☐ + 3 7

Fill in the Missing Number.

1) □ + 4 = 7	2) □ + 2 = 4	3) □ + 2 = 3
4) □ + 4 = 8	5) □ + 2 = 5	6) □ + 3 = 8
7) □ + 4 = 5	8) □ + 3 = 8	9) □ + 1 = 4
10) □ + 4 = 8	11) □ + 5 = 6	12) □ + 5 = 8
13) □ + 4 = 6	14) □ + 4 = 7	15) □ + 5 = 7

Fill in the Missing Number.

1) $\square + 2 = 3$

2) $\square + 3 = 4$

3) $\square + 1 = 3$

4) $\square + 5 = 7$

5) $\square + 2 = 7$

6) $\square + 3 = 8$

7) $\square + 5 = 10$

8) $\square + 2 = 7$

9) $\square + 5 = 6$

10) $\square + 5 = 6$

11) $\square + 1 = 5$

12) $\square + 4 = 8$

13) $\square + 5 = 10$

14) $\square + 5 = 7$

15) $\square + 5 = 10$

Fill in the Missing Number.

1) ☐ + 2 = 5

2) ☐ + 3 = 5

3) ☐ + 4 = 5

4) ☐ + 4 = 5

5) ☐ + 4 = 7

6) ☐ + 4 = 8

7) ☐ + 5 = 8

8) ☐ + 3 = 8

9) ☐ + 1 = 3

10) ☐ + 1 = 6

11) ☐ + 5 = 7

12) ☐ + 1 = 3

13) ☐ + 3 = 4

14) ☐ + 4 = 7

15) ☐ + 4 = 6

Fill in the Missing Number.

1) □ + 2 = 7

2) □ + 2 = 4

3) □ + 5 = 8

4) □ + 2 = 5

5) □ + 1 = 6

6) □ + 1 = 6

7) □ + 4 = 8

8) □ + 3 = 8

9) □ + 3 = 4

10) □ + 3 = 4

11) □ + 3 = 8

12) □ + 2 = 3

13) □ + 1 = 3

14) □ + 5 = 10

15) □ + 1 = 3

Fill in the Missing Number.

1) $\square + 5 = 9$

2) $\square + 5 = 10$

3) $\square + 3 = 6$

4) $\square + 5 = 10$

5) $\square + 4 = 7$

6) $\square + 2 = 3$

7) $\square + 2 = 3$

8) $\square + 2 = 4$

9) $\square + 3 = 6$

10) $\square + 1 = 2$

11) $\square + 1 = 3$

12) $\square + 2 = 3$

13) $\square + 2 = 4$

14) $\square + 4 = 5$

15) $\square + 4 = 8$

Fill in the Missing Number.

1) □ + 2 = 5

2) □ + 2 = 7

3) □ + 2 = 3

4) □ + 1 = 4

5) □ + 4 = 8

6) □ + 5 = 10

7) □ + 2 = 7

8) □ + 3 = 7

9) □ + 5 = 8

10) □ + 3 = 7

11) □ + 4 = 9

12) □ + 4 = 5

13) □ + 5 = 7

14) □ + 4 = 5

15) □ + 5 = 9

Fill in the Missing Number.

1) $3 + \square = 8$

2) $5 + \square = 7$

3) $4 + \square = 7$

4) $2 + \square = 5$

5) $5 + \square = 10$

6) $2 + \square = 5$

7) $2 + \square = 4$

8) $2 + \square = 6$

9) $1 + \square = 5$

10) $3 + \square = 6$

11) $4 + \square = 5$

12) $3 + \square = 5$

13) $1 + \square = 5$

14) $5 + \square = 8$

15) $5 + \square = 9$

Fill in the Missing Number.

1) $1 + \square = 2$

2) $2 + \square = 6$

3) $1 + \square = 3$

4) $5 + \square = 10$

5) $1 + \square = 6$

6) $1 + \square = 4$

7) $5 + \square = 10$

8) $3 + \square = 4$

9) $5 + \square = 9$

10) $5 + \square = 9$

11) $4 + \square = 6$

12) $2 + \square = 5$

13) $2 + \square = 7$

14) $1 + \square = 5$

15) $5 + \square = 8$

Fill in the Missing Number.

1) $4 + \square = 8$

2) $5 + \square = 7$

3) $5 + \square = 8$

4) $1 + \square = 2$

5) $2 + \square = 7$

6) $3 + \square = 7$

7) $3 + \square = 7$

8) $1 + \square = 3$

9) $3 + \square = 7$

10) $4 + \square = 5$

11) $1 + \square = 3$

12) $5 + \square = 10$

13) $1 + \square = 4$

14) $5 + \square = 6$

15) $3 + \square = 4$

Fill in the Missing Number.

1) $4 + \square = 5$

2) $1 + \square = 4$

3) $2 + \square = 5$

4) $5 + \square = 10$

5) $4 + \square = 6$

6) $5 + \square = 8$

7) $2 + \square = 3$

8) $1 + \square = 2$

9) $2 + \square = 4$

10) $1 + \square = 4$

11) $1 + \square = 4$

12) $1 + \square = 2$

13) $3 + \square = 5$

14) $1 + \square = 2$

15) $3 + \square = 5$

Fill in the Missing Number.

1) $\begin{array}{r} 1 \\ + \square \\ \hline 5 \end{array}$

2) $\begin{array}{r} 2 \\ + \square \\ \hline 7 \end{array}$

3) $\begin{array}{r} 5 \\ + \square \\ \hline 8 \end{array}$

4) $\begin{array}{r} 3 \\ + \square \\ \hline 4 \end{array}$

5) $\begin{array}{r} 2 \\ + \square \\ \hline 7 \end{array}$

6) $\begin{array}{r} 3 \\ + \square \\ \hline 7 \end{array}$

7) $\begin{array}{r} 4 \\ + \square \\ \hline 9 \end{array}$

8) $\begin{array}{r} 5 \\ + \square \\ \hline 8 \end{array}$

9) $\begin{array}{r} 4 \\ + \square \\ \hline 6 \end{array}$

10) $\begin{array}{r} 3 \\ + \square \\ \hline 4 \end{array}$

11) $\begin{array}{r} 4 \\ + \square \\ \hline 7 \end{array}$

12) $\begin{array}{r} 2 \\ + \square \\ \hline 4 \end{array}$

13) $\begin{array}{r} 2 \\ + \square \\ \hline 3 \end{array}$

14) $\begin{array}{r} 2 \\ + \square \\ \hline 5 \end{array}$

15) $\begin{array}{r} 3 \\ + \square \\ \hline 4 \end{array}$

Fill in the Missing Number.

1) $\begin{array}{r} 1 \\ + \square \\ \hline 5 \end{array}$	2) $\begin{array}{r} 2 \\ + \square \\ \hline 3 \end{array}$	3) $\begin{array}{r} 2 \\ + \square \\ \hline 4 \end{array}$
4) $\begin{array}{r} 4 \\ + \square \\ \hline 6 \end{array}$	5) $\begin{array}{r} 5 \\ + \square \\ \hline 9 \end{array}$	6) $\begin{array}{r} 4 \\ + \square \\ \hline 5 \end{array}$
7) $\begin{array}{r} 1 \\ + \square \\ \hline 4 \end{array}$	8) $\begin{array}{r} 2 \\ + \square \\ \hline 4 \end{array}$	9) $\begin{array}{r} 2 \\ + \square \\ \hline 5 \end{array}$
10) $\begin{array}{r} 4 \\ + \square \\ \hline 7 \end{array}$	11) $\begin{array}{r} 5 \\ + \square \\ \hline 9 \end{array}$	12) $\begin{array}{r} 4 \\ + \square \\ \hline 9 \end{array}$
13) $\begin{array}{r} 3 \\ + \square \\ \hline 7 \end{array}$	14) $\begin{array}{r} 3 \\ + \square \\ \hline 6 \end{array}$	15) $\begin{array}{r} 3 \\ + \square \\ \hline 7 \end{array}$

Fill in the Missing Number.

1) $1 + \square = 4$

2) $3 + \square = 7$

3) $1 + \square = 2$

4) $1 + \square = 5$

5) $5 + \square = 9$

6) $2 + \square = 5$

7) $4 + \square = 7$

8) $3 + \square = 5$

9) $3 + \square = 4$

10) $3 + \square = 8$

11) $1 + \square = 4$

12) $3 + \square = 7$

13) $2 + \square = 5$

14) $2 + \square = 4$

15) $1 + \square = 3$

Fill in the Missing Number.

1) $\begin{array}{r} 1 \\ + \square \\ \hline 4 \end{array}$
2) $\begin{array}{r} 5 \\ + \square \\ \hline 8 \end{array}$
3) $\begin{array}{r} 3 \\ + \square \\ \hline 5 \end{array}$

4) $\begin{array}{r} 2 \\ + \square \\ \hline 5 \end{array}$
5) $\begin{array}{r} 2 \\ + \square \\ \hline 5 \end{array}$
6) $\begin{array}{r} 4 \\ + \square \\ \hline 8 \end{array}$

7) $\begin{array}{r} 1 \\ + \square \\ \hline 6 \end{array}$
8) $\begin{array}{r} 3 \\ + \square \\ \hline 4 \end{array}$
9) $\begin{array}{r} 2 \\ + \square \\ \hline 3 \end{array}$

10) $\begin{array}{r} 1 \\ + \square \\ \hline 6 \end{array}$
11) $\begin{array}{r} 4 \\ + \square \\ \hline 5 \end{array}$
12) $\begin{array}{r} 4 \\ + \square \\ \hline 8 \end{array}$

13) $\begin{array}{r} 5 \\ + \square \\ \hline 7 \end{array}$
14) $\begin{array}{r} 2 \\ + \square \\ \hline 5 \end{array}$
15) $\begin{array}{r} 2 \\ + \square \\ \hline 3 \end{array}$

Find the Sum.

1) 5 + 5 = ☐

2) 1 + 4 = ☐

3) 2 + 2 = ☐

4) 3 + 1 = ☐

5) 4 + 1 = ☐

6) 1 + 1 = ☐

7) 3 + 5 = ☐

8) 3 + 2 = ☐

9) 4 + 4 = ☐

10) 2 + 5 = ☐

11) 4 + 1 = ☐

12) 1 + 1 = ☐

13) 2 + 1 = ☐

14) 3 + 1 = ☐

15) 4 + 5 = ☐

Find the Sum.

1) $\begin{array}{r} 4 \\ +\ 5 \\ \hline \square \end{array}$	2) $\begin{array}{r} 5 \\ +\ 3 \\ \hline \square \end{array}$	3) $\begin{array}{r} 4 \\ +\ 5 \\ \hline \square \end{array}$
4) $\begin{array}{r} 5 \\ +\ 1 \\ \hline \square \end{array}$	5) $\begin{array}{r} 2 \\ +\ 1 \\ \hline \square \end{array}$	6) $\begin{array}{r} 5 \\ +\ 1 \\ \hline \square \end{array}$
7) $\begin{array}{r} 1 \\ +\ 2 \\ \hline \square \end{array}$	8) $\begin{array}{r} 4 \\ +\ 5 \\ \hline \square \end{array}$	9) $\begin{array}{r} 3 \\ +\ 4 \\ \hline \square \end{array}$
10) $\begin{array}{r} 2 \\ +\ 2 \\ \hline \square \end{array}$	11) $\begin{array}{r} 3 \\ +\ 2 \\ \hline \square \end{array}$	12) $\begin{array}{r} 3 \\ +\ 2 \\ \hline \square \end{array}$
13) $\begin{array}{r} 3 \\ +\ 5 \\ \hline \square \end{array}$	14) $\begin{array}{r} 4 \\ +\ 2 \\ \hline \square \end{array}$	15) $\begin{array}{r} 3 \\ +\ 1 \\ \hline \square \end{array}$

Find the Sum.

1) $\begin{array}{r} 2 \\ +\ 3 \\ \hline \square \end{array}$	2) $\begin{array}{r} 5 \\ +\ 3 \\ \hline \square \end{array}$	3) $\begin{array}{r} 1 \\ +\ 5 \\ \hline \square \end{array}$
4) $\begin{array}{r} 4 \\ +\ 1 \\ \hline \square \end{array}$	5) $\begin{array}{r} 3 \\ +\ 3 \\ \hline \square \end{array}$	6) $\begin{array}{r} 3 \\ +\ 5 \\ \hline \square \end{array}$
7) $\begin{array}{r} 1 \\ +\ 3 \\ \hline \square \end{array}$	8) $\begin{array}{r} 3 \\ +\ 2 \\ \hline \square \end{array}$	9) $\begin{array}{r} 1 \\ +\ 3 \\ \hline \square \end{array}$
10) $\begin{array}{r} 3 \\ +\ 5 \\ \hline \square \end{array}$	11) $\begin{array}{r} 1 \\ +\ 2 \\ \hline \square \end{array}$	12) $\begin{array}{r} 5 \\ +\ 4 \\ \hline \square \end{array}$
13) $\begin{array}{r} 5 \\ +\ 5 \\ \hline \square \end{array}$	14) $\begin{array}{r} 5 \\ +\ 3 \\ \hline \square \end{array}$	15) $\begin{array}{r} 3 \\ +\ 3 \\ \hline \square \end{array}$

Find the Sum.

1) $1 + 2 =$ [ ]

2) $2 + 5 =$ [ ]

3) $2 + 1 =$ [ ]

4) $1 + 2 =$ [ ]

5) $2 + 1 =$ [ ]

6) $3 + 1 =$ [ ]

7) $2 + 1 =$ [ ]

8) $5 + 1 =$ [ ]

9) $5 + 3 =$ [ ]

10) $5 + 3 =$ [ ]

11) $4 + 2 =$ [ ]

12) $2 + 2 =$ [ ]

13) $4 + 2 =$ [ ]

14) $5 + 2 =$ [ ]

15) $1 + 3 =$ [ ]

Find the Sum.

1) $5 + 4$	2) $4 + 5$	3) $5 + 5$
4) $1 + 1$	5) $1 + 3$	6) $1 + 3$
7) $4 + 2$	8) $1 + 5$	9) $1 + 3$
10) $1 + 3$	11) $5 + 1$	12) $5 + 3$
13) $1 + 2$	14) $3 + 4$	15) $3 + 1$

Find the Sum.

1) $\begin{array}{r} 5 \\ +\ 3 \\ \hline \square \end{array}$	2) $\begin{array}{r} 4 \\ +\ 3 \\ \hline \square \end{array}$	3) $\begin{array}{r} 3 \\ +\ 5 \\ \hline \square \end{array}$
4) $\begin{array}{r} 5 \\ +\ 2 \\ \hline \square \end{array}$	5) $\begin{array}{r} 2 \\ +\ 2 \\ \hline \square \end{array}$	6) $\begin{array}{r} 4 \\ +\ 3 \\ \hline \square \end{array}$
7) $\begin{array}{r} 2 \\ +\ 2 \\ \hline \square \end{array}$	8) $\begin{array}{r} 4 \\ +\ 5 \\ \hline \square \end{array}$	9) $\begin{array}{r} 1 \\ +\ 4 \\ \hline \square \end{array}$
10) $\begin{array}{r} 4 \\ +\ 3 \\ \hline \square \end{array}$	11) $\begin{array}{r} 5 \\ +\ 2 \\ \hline \square \end{array}$	12) $\begin{array}{r} 4 \\ +\ 5 \\ \hline \square \end{array}$
13) $\begin{array}{r} 1 \\ +\ 5 \\ \hline \square \end{array}$	14) $\begin{array}{r} 1 \\ +\ 2 \\ \hline \square \end{array}$	15) $\begin{array}{r} 3 \\ +\ 1 \\ \hline \square \end{array}$

Find the Sum.

1) 2 + 4 = ☐

2) 2 + 4 = ☐

3) 2 + 1 = ☐

4) 5 + 4 = ☐

5) 5 + 5 = ☐

6) 5 + 4 = ☐

7) 3 + 5 = ☐

8) 5 + 4 = ☐

9) 2 + 4 = ☐

10) 4 + 4 = ☐

11) 1 + 1 = ☐

12) 2 + 3 = ☐

13) 4 + 2 = ☐

14) 4 + 5 = ☐

15) 2 + 1 = ☐

Find the Sum.

1) $4 + 4 = \square$

2) $4 + 1 = \square$

3) $5 + 3 = \square$

4) $2 + 1 = \square$

5) $1 + 3 = \square$

6) $2 + 4 = \square$

7) $3 + 4 = \square$

8) $1 + 1 = \square$

9) $5 + 3 = \square$

10) $2 + 4 = \square$

11) $4 + 1 = \square$

12) $3 + 5 = \square$

13) $5 + 2 = \square$

14) $2 + 5 = \square$

15) $4 + 3 = \square$

Page 1, Item 1:
(1)2 (2)5 (3)3 (4)5 (5)1 (6)3 (7)1 (8)4 (9)2 (10)3 (11)3 (12)4
Page 2, Item 1:
(1)4 (2)5 (3)5 (4)5 (5)1 (6)4 (7)4 (8)5 (9)1 (10)2 (11)3 (12)4
Page 3, Item 1:
(1)3 (2)5 (3)2 (4)2 (5)5 (6)4 (7)3 (8)4 (9)3 (10)3 (11)3 (12)5
Page 4, Item 1:
(1)4 (2)3 (3)3 (4)1 (5)4 (6)2 (7)1 (8)1 (9)3 (10)3 (11)2 (12)4
Page 5, Item 1:
(1)4 (2)5 (3)1 (4)1 (5)3 (6)1 (7)5 (8)2 (9)3 (10)3 (11)2 (12)2
Page 6, Item 1:
(1)5 (2)5 (3)5 (4)5 (5)3 (6)2 (7)1 (8)1 (9)1 (10)3 (11)5 (12)2
Page 7, Item 1:
(1)4 (2)1 (3)1 (4)3 (5)4 (6)5 (7)4 (8)5 (9)1 (10)1 (11)1 (12)1
Page 8, Item 1:
(1)5 (2)3 (3)3 (4)2 (5)4 (6)4 (7)5 (8)2 (9)4 (10)3 (11)1 (12)4
Page 9, Item 1:
(1)2 (2)5 (3)2 (4)1 (5)2 (6)4 (7)4 (8)4 (9)3 (10)3 (11)5 (12)2
Page 10, Item 1:
(1)5 (2)3 (3)2 (4)4 (5)4 (6)2 (7)3 (8)5 (9)3 (10)4 (11)3 (12)4
Page 11, Item 1:
(1)1 (2)4 (3)3 (4)3 (5)1 (6)1 (7)1 (8)3 (9)2 (10)2 (11)5 (12)5
Page 12, Item 1:
(1)2 (2)4 (3)4 (4)1 (5)4 (6)5 (7)1 (8)4 (9)5 (10)3 (11)5 (12)4
Page 13, Item 1:
(1)4 (2)3 (3)2 (4)5 (5)4 (6)3 (7)3 (8)3 (9)5 (10)4 (11)5 (12)3
Page 14, Item 1:
(1)1 (2)5 (3)3 (4)4 (5)5 (6)2 (7)2 (8)4 (9)4 (10)5 (11)4 (12)3
Page 15, Item 1:
(1)2 (2)3 (3)1 (4)1 (5)4 (6)1 (7)5 (8)5 (9)5 (10)4 (11)3 (12)4
Page 16, Item 1:
(1)3 (2)1 (3)5 (4)2 (5)2 (6)5 (7)3 (8)1 (9)4 (10)2 (11)5 (12)3
Page 17, Item 1:
(1)6 (2)9 (3)9 (4)5 (5)7 (6)5 (7)6 (8)6 (9)5 (10)10 (11)4 (12)8
Page 18, Item 1:
(1)8 (2)7 (3)8 (4)5 (5)4 (6)8 (7)8 (8)6 (9)6 (10)3 (11)9 (12)5
Page 19, Item 1:
(1)6 (2)6 (3)10 (4)4 (5)8 (6)3 (7)8 (8)6 (9)6 (10)9 (11)7 (12)8
Page 20, Item 1:
(1)9 (2)5 (3)9 (4)9 (5)5 (6)6 (7)7 (8)9 (9)8 (10)5 (11)5 (12)10
Page 21, Item 1:
(1)5 (2)5 (3)8 (4)7 (5)2 (6)8 (7)6 (8)6 (9)4 (10)5 (11)5 (12)9
Page 22, Item 1:
(1)3 (2)8 (3)7 (4)3 (5)6 (6)7 (7)6 (8)8 (9)8 (10)6 (11)6 (12)6

Page 23, Item 1:
(1)8 (2)6 (3)5 (4)6 (5)6 (6)6 (7)2 (8)5 (9)6
(10)9 (11)10 (12)9
Page 24, Item 1:
(1)2 (2)8 (3)8 (4)3 (5)6 (6)6 (7)6 (8)5 (9)7
(10)4 (11)5 (12)8
Page 25, Item 1:
(1)4 (2)3 (3)5 (4)5 (5)4 (6)4 (7)4 (8)1 (9)1
(10)1 (11)5 (12)1 (13)4 (14)4 (15)5
Page 26, Item 1:
(1)2 (2)5 (3)3 (4)3 (5)2 (6)4 (7)1 (8)3 (9)2
(10)4 (11)1 (12)5 (13)1 (14)2 (15)4
Page 27, Item 1:
(1)3 (2)2 (3)1 (4)4 (5)3 (6)5 (7)1 (8)5 (9)3
(10)4 (11)1 (12)3 (13)2 (14)3 (15)2
Page 28, Item 1:
(1)1 (2)1 (3)2 (4)2 (5)5 (6)5 (7)5 (8)5 (9)1
(10)1 (11)4 (12)4 (13)5 (14)2 (15)5
Page 29, Item 1:
(1)3 (2)2 (3)1 (4)1 (5)3 (6)4 (7)3 (8)5 (9)2
(10)5 (11)2 (12)2 (13)1 (14)3 (15)2
Page 30, Item 1:
(1)5 (2)2 (3)3 (4)3 (5)5 (6)5 (7)4 (8)5 (9)1
(10)1 (11)5 (12)1 (13)2 (14)5 (15)2
Page 31, Item 1:
(1)4 (2)5 (3)3 (4)5 (5)3 (6)1 (7)1 (8)2 (9)3
(10)1 (11)2 (12)1 (13)2 (14)1 (15)4
Page 32, Item 1:
(1)3 (2)5 (3)1 (4)3 (5)4 (6)5 (7)5 (8)4 (9)3
(10)4 (11)5 (12)1 (13)2 (14)1 (15)4
Page 33, Item 1:
(1)5 (2)2 (3)3 (4)3 (5)5 (6)3 (7)2 (8)4 (9)4
(10)3 (11)1 (12)2 (13)4 (14)3 (15)4
Page 34, Item 1:
(1)1 (2)4 (3)2 (4)5 (5)5 (6)3 (7)5 (8)1 (9)4
(10)4 (11)2 (12)3 (13)5 (14)4 (15)3
Page 35, Item 1:
(1)4 (2)2 (3)3 (4)1 (5)5 (6)4 (7)4 (8)2 (9)4
(10)1 (11)2 (12)5 (13)3 (14)1 (15)1
Page 36, Item 1:
(1)1 (2)3 (3)3 (4)5 (5)2 (6)3 (7)1 (8)1 (9)2
(10)3 (11)3 (12)1 (13)2 (14)1 (15)2
Page 37, Item 1:
(1)4 (2)5 (3)3 (4)1 (5)5 (6)4 (7)5 (8)3 (9)2
(10)1 (11)3 (12)2 (13)1 (14)3 (15)1
Page 38, Item 1:
(1)4 (2)1 (3)2 (4)2 (5)4 (6)1 (7)3 (8)2 (9)3
(10)3 (11)4 (12)5 (13)4 (14)3 (15)4
Page 39, Item 1:
(1)3 (2)4 (3)1 (4)4 (5)4 (6)3 (7)3 (8)2 (9)1
(10)5 (11)3 (12)4 (13)3 (14)2 (15)2
Page 40, Item 1:
(1)3 (2)3 (3)2 (4)3 (5)3 (6)4 (7)5 (8)1 (9)1
(10)5 (11)1 (12)4 (13)2 (14)3 (15)1
Page 41, Item 1:
(1)10 (2)5 (3)4 (4)4 (5)5 (6)2 (7)8 (8)5 (9)8
(10)7 (11)5 (12)2 (13)3 (14)4 (15)9
Page 42, Item 1:
(1)9 (2)8 (3)9 (4)6 (5)3 (6)6 (7)3 (8)9 (9)7
(10)4 (11)5 (12)5 (13)8 (14)6 (15)4
Page 43, Item 1:
(1)5 (2)8 (3)6 (4)5 (5)6 (6)8 (7)4 (8)5 (9)4
(10)8 (11)3 (12)9 (13)10 (14)8 (15)6
Page 44, Item 1:
(1)3 (2)7 (3)3 (4)3 (5)3 (6)4 (7)3 (8)6 (9)8
(10)8 (11)6 (12)4 (13)6 (14)7 (15)4

Page 45, Item 1:
(1)9 (2)9 (3)10 (4)2 (5)4 (6)4 (7)6 (8)6 (9)4
(10)4 (11)6 (12)8 (13)3 (14)7 (15)4
Page 46, Item 1:
(1)8 (2)7 (3)8 (4)7 (5)4 (6)7 (7)4 (8)9 (9)5
(10)7 (11)7 (12)9 (13)6 (14)3 (15)4
Page 47, Item 1:
(1)6 (2)6 (3)3 (4)9 (5)10 (6)9 (7)8 (8)9 (9)6
(10)8 (11)2 (12)5 (13)6 (14)9 (15)3
Page 48, Item 1:
(1)8 (2)5 (3)8 (4)3 (5)4 (6)6 (7)7 (8)2 (9)8
(10)6 (11)5 (12)8 (13)7 (14)7 (15)7

Find More From

LEARN AND GROW

@

TKTCollection Publishing

on Amazon

or

TKTCOLLECTION.COM

IF YOU LOVE THIS BOOK, PLEASE CONSIDER LEAVING A REVIEW ON AMAZON. THERE IS NO OBLIGATION AND IT DOES TAKE A FEW MINUTES, BUT ALL FEEDBACK IS VERY MUCH APPRECIATED.

THANK YOU!

Made in United States
North Haven, CT
07 November 2026

21928347R00033